MATHEMATICS
ONE DAY
[Practice Books]

Sujit Kumar Mishra

Mathematics One Day

By Sujit Kumar Mishra

Blogs : https://sujitvital89.blogspot.com

Copywriter © 2020 Sujit Kumar Mishra

All Right Reserved. No Part of this publication may be reproduced or distributed in any from or by any means or store in database or retrieval system or device. Without the period written permission of publisher, with the exception the program listing may be entered, stored and executed in a computer system, but may they not be reproduced for publication.

This book is dedicated to all mathematics candidates, Student, who is preparing for a comparative examination. This book is a collection of all in one.

TABLE OF CONTENTS

PART # 1

SECTION 1: NUMBERS SERIES
NUMERICAL ABILITY TEST
NUMERICAL ABILITY TEST ANSWER
NUMERICAL ABILITY TEST OBJECTIVE TYPE QUESTIONS
NUMERICAL ABILITY TEST OBJECTIVE TYPE ANSWERS
NUMBER SEQUENCE
OBJECTIVE TYPE NUMBER SERIES
OBJECTIVE TYPE NUMBER SERIES ANSWER
WRONG NUMBER SERIES

SECTION 2: PERCENTAGE
PERCENTAGE BASIC
PERCENTAGE OBJECTIVE TYPES QUESTIONS
PERCENTAGE OBJECTIVE TYPES QUESTION ANSWERS

SECTION 3: PROFIT & LOSS
PROFIT & LOSS PRACTICE
PROFIT & LOSS OBJECTIVE TYPE QUESTIONS
PROFIT & LOSS OBJECTIVE TYPE QUESTIONS ANSWERS

SECTION 3: AVERAGE
AVERAGE PRACTICE

Section 4: Time & distance
Time Speed & Distance……………………………………………….. 92

Section 5: Algebra
Basic Algebra……………………………………………………………102

Section 6: Time & Work
Time and work series……………………………………………………108

Section 7: Ratio & Proportion
Ratio & Proration Series………………………………………………….115

Section 8: SI & CI
Simple & Compound Interest……………………………………….121

PART # 2……………………………………………………………….130

Tricks of Number………………………………………………. 130
Formulas Collection # 1……………………………………… 131
Formula Collection # 2 …………………………………….. 138

PREFACE

The one day mathematic is part of interview preparation for the Comparative examination. This book is also beneficial for everyone who is prepared for any competitive examination like multinational Company, government, semi-government, Bank, Insurance Sector & Relevant Organization. you can learn easily mathematics. In this, you can also learn the collection of most frequently ask formulas.

SECTION 1

NUMBERS SERIES

"Be the change that you want to see in the world."
--mahatma Gandhi

NUMERICAL ABILITY TESTS

1) **Arithmetic Question Part One Solve the question as given below?**

a) 335 + 152 =?

b) 189 x 125 =?

c) 1354 – 1543 =?

d) 1235/25 =?

e) 4569 + 156 + 789 =?

f) 125-80=?

g) 789-854=?

h) 98752-159-159=?

i) 789 x 12-15?

j) 987-155-25=?

k) 148-89-147=?

l) 9859+458-123=?

m) 85467-214-157+456=?

n) 987-254+224+458+20=?

o) 9654+354-845+78=?

p) 3217-84-25=?

q) 9654-154+78=?

r) 95175-3214+7854=?

s) 854 x 12 + 155=?

t) 456 x 65 +125 =?

u) 6324-126-78+30 x 30 =?

v) 3511+125 - (788 x 15) =?

w) 952-224+314 x 99 =?

x) 3214-7898+859 x 789 =?

y) 3265 – 32 x 965 + 92=?

z) 7524+664-54 x 60+89 =?

aa) 124+2342-124 x 65 / 12 =?

bb) 125-210+89 x 9 /10 =?

cc) 654+745-22 x 75 + 6=?

dd) 954/32 + 654 x 65 -3=?

ee) 2813+7897+84-489=?

ff) 920/120 x 30 + 60=?

A	487	Q	9578	GG	3570
B	23625	R	99815	HH	53463
C	-189	S	10403	II	892.57
D	49.4	T	29765	JJ	17275
E	5514	U	7020	KK	5511.16
F	45	V	-8184	LL	10566
G	-65	W	31814		
H	98434	X	673067		
I	9453	Y	-27532		

gg) 765+78-85 x 34 =?

hh) 954 x 56 – 56 +95 =?

ii) 545-66+7858/19=?

jj) 965 x 18 -159 + 64=?

kk) 4502+1024-950/64=?

ll) 850-654+305 x 34=?

J	807	Z	5037		
K	-88	AA	1794		
L	10194	BB	-4.9		
M	85552	CC	-245		
N	1435	DD	42536.81		
O	9241	EE	10305		
P	3180	FF	290		

NUMERICAL ABILITY TEST ANSWER

Rule of BODMOS

The BODMOS Meaning Bracket, Of, Division, Multiplication, Addition and Subtraction.

Numerical Ability Test Objective Type Questions

2) Arithmetic Question Part One Solve the question as given below & Choose Appropriate Answers?

   i) 150+30+89 =

a) 289	b) 269

| c) 300 | d) 358 |

ii) 189+578+45 =

| a) 812 | b) 813 |
| c) 815 | d) 890 |

iii) 987+205+489=

| a) 1682 | b) 1658 |
| c) 1681 | d) 1695 |

iv) 1598+7852+4569=

| a) 14019 | b) 1408 |
| c) 1420 | d) 1508 |

v) 203+4885+30=

| a) 5119 | b) 1618 |
| c) 5118 | d) 5119 |

vi) 89654+2348+158+60=

| a) 92654 | b)32566 |
| c)92645 | d)92261 |

    vii)   789+60+5005+1111=

| a) 7065 | b)7064 |
| c)6965 | d)7065 |

    viii)   60+444+4587+89=

| a) 5190 | b)5180 |
| c)5195 | d)6195 |

    ix)   20+602-20=

| a) 605 | b)602 |
| c)603 | d)604 |

    x)   600+450-99=

| a) 952 | b)953 |

c)951	d)900

xi) 802-640+105=

a) 267	b)629
c)268	d)270

xii) 2364+2564-89=

a) 4838	b)4840
c)4839	d) 4856

xiii) 3250-90+1598=

a) 4849	b) 4749
c)4879	d)4750

xiv) 1530-892+30+148=

a) 3215	b)800

| c)818 | d)816 |

xv)   320+89-158+1478=

| a) 1730 | b)1729 |
| c) 1771 | d)1483 |

xvi)   -2563+20+1000+555=

| a) -888 | b)988 |
| c) -988 | d) -990 |

xvii)   32+123-20 x 18 =

| a) 320 | b) -205 |
| c) 205 | d) 300 |

xviii)   321+62 x 159 -2500=

| a) 7679 | b) 7634 |
| c) 7680 | d) 8006 |

xix)   60 x 14 + 999 – 222=

| a) 1678 | b)1764 |

| c)1617 | |

xx)   1598+654-354 x 95=

| a) -31378 | b)31378 |
| c) 32546 | d) -31379 |

xxi)   324-154+245 x 55=

| a) 13645 | b)13545 |
| c)13550 | 13646 |

xxii)   33333+1111+222-999=

| a) 333649 | b)33466 |
| c)33667 | d)33690 |

xxiii)   456789+12456-9990-459200=

| a) 55 | b) 60 |
| c)56 | d)133566 |

xxiv) 5649-9000-111-39=

a) -3505	b) 6541
c) 3505	d) -3501

xxv) 570 x 18 + 999 – 1111=

a) 10148	b) 11149
c) 11113	d) 10135

xxvi) 654-500+12 x 89=

a) 1223	b) 1222
c) 1322	d) 1232

xxvii) 425/12 + 20 -15=

a) 40.41	b) 44.22
c) 45.02	d) 55.87

xxviii) 888/22 + 89 – 159 x 5=

| a) -665.64 | b) -665.99 |
| c) 668 | d) -661.20 |

xxiv) 654 + 125/25 – 99=

| a) 567 | b) 570 |
| c) 560 | d) 590 |

xxv) 654/32 x 9 – 8 + 444=

| a) 612 | b) 621 |
| c) 619.93 | d) 625.29 |

xxvi) 999+9999+99-111=

| a) 10986 | b) 10990 |
| c) 10987 | d) 10999 |

xxvii) 8888+88-8 x 5+450/45=

| a) 9020 | b) 9017 |

| c)9015 | d)9018 |

xxviii) 20 % of 300=

| a) 66 | b)60 |
| c)65 | d)150 |

xxix) 85% of 999=

| a) 850 | b) 849.85 |
| c) 900 | d) 990 |

xxx) 1054+900(40% of 600) =

| a) 2195 | b) 2194 |
| c) 2224 | d) 9654 |

xxxi) 1000-99(75% of 2000) =

| a) 2401 | b)2403 |

| c)2409 | d)2405 |

NUMERICAL ABILITY TEST OBJECTIVE TYPE ANSWERS

i)	B	ii)	A	iii)	C
iv)	A	v)	C	vi)	D
vii)	C	viii)	B	ix)	B
x)	C	xi)	A	xii)	C

xiii)	B	xiv)	D	xv)	B
xvi)	C	xvii)	B	xviii)	C
xiv)	C	xx)	A	xxi)	A
xxii)	C	xxiii)	A	xxiv)	D
xxv)	A	xxvi)	B	xxvii)	A
xxviii)	A	xxiv)	C	xxv)	C
xxvi)	A	xxvii)	D	xxviii)	B
xxix)	B	xxx)	B	xxxi)	A

Number Sequence

The Number Sequence is list of Number which is connected each other with his linked list or node as per possible rules. you can work out the next numbers in the sequence.

1) **Adding Number Example for Number Series fill the blanks:**

a) 5,7,9,11,13,14,15, _____ .

In this series each number have adding +2 Example:

5+2= 7

7+2=9

9+2=11

11+2=13

13+2=15

15+2=_17_

The Answer will be 17 in this blanks.

2) **Subtracting number Example for Number Series fill the blanks. (The common difference could also be negative)**

b) **25,20,15,10, ___**

In this series each number have subtracting number by -5 with decrease series Example:

Example # 1

25-5=20

20-5=15

15-5=10

10-5=_5_

The Answer will be 5 in this blanks:

Example # II

    c) ___, 100,120,130,140.

140-10=130

130-10=120

120-10=100

100-10=**_90_**

The Answer will be 90 in this blanks.

    3) **The multiply by same value of each time in series example:**

Example # 1

    d) 1,3,2,9,3,27,4,___.

In this function we can break as given below:

Series One: 1,2,3,4---------------- Leave One

Series Two: 3,9,27, ___ ---------Work

1 x 3 = 3

3 x 3 = 9

9 x 3 = 27

27 x 3 = **_81_**

The Answer will be 81 in this blanks.

Example # II

    e)   10, 5, 2.5, 1.25, 0.625, 0.3125, ____

    4) In this sequence starts at 10 and has a common ratio of 0.5 (a half). On this series the pattern is continued by multiplying by 0.5 each time.

10 x 0.5 = 2.5

2.5 x 0.5 = 1.25

1.25 x 0.5 = 0.625

0.625 x 0.5 = 0.325

0.325 x 0.5 = **_0.1625_**

The Answer will be 0.1625 in this blanks.

5) **How to used Triangular number in Number Series Example as given below:**

f) 1,4,8,13,19,26,___,34,43.

1 + 1 = 4

4 + 4 = 8

8 + 5 = 13

13 + 6 = 19

19 + 7 = 26

26 + 8 = **_34_**

34 + 9 = 43

The Answer will be 34 in this blanks.

6) How to calculate Square & cube Number in Series as given below:

Example # 1:> (Square)

g) 1,4,9,16,25, ___,49,64.

1 x 1 = 1
2 x 2 = 4
3 x 3 = 9
4 x 4 = 16
5 x 5 = 25
6 x 6 = 36
7 x 7 = **_49_**
8 x 8 = 64

The Answer will be 49 in this blanks.

Example # II:> (Cube)

h) 1,8,27,64,125___.

1 x 1 x 1 = 1
2 x 2 x 2 = 8
3 x 3 x 3 = 27
4 x 4 x 4 = 64

5 x 5 x 5 = 125

6 x 6 x 6 = **_216_**

The Answer will be 216 in this blanks.

7) **How to calculate Fibonacci Series in Number Sequence Example as given below:**

   i) 0,1,1,2,3,5,8,13, ____

1 + 1 = 2

3 + 1 = 3

3 + 2 = 5

5 + 3 = 8

8 + 5 = 13

13 + 8 = **_21_**

The Answer will be 21 in this blanks.

8) **How to calculate Division number in Series Example as given below:**

   j) 2,5,7,9___.

10 / 5 = 2

25 / 5 = 5

$35 / 5 = 7$

$45 / 5 = 9$

$55 / 5 = \underline{\boldsymbol{\mathit{11}}}$

The Answer will be 11 in this blanks.

Objective Type Number Series

Find out appropriate answer of series & find out what number should come next?

1) 5,8,11,14,17,20,23, _____ ?

| a) 27 | b) 26 |

| c) 28 | d) 30 |

2) 10,15,20,25,30,__,40,45,50.

| a) 35 | b) 40 |
| c) 36 | d) 37 |

3) 123,130,137,144,151,__?

| a) 159 | b) 160 |
| c) 158 | d) 159 |

4) 1234,1238,1242,1246,1250,__?

| a) 1254 | b) 1255 |
| c) 1256 | d) 1258 |

5) 90,84,76,72,66,__?

| a) 59 | b) 60 |
| c) 61 | d) 62 |

6) 100,98,96,___,92,90,88.

| a) 94 | b) 95 |

| c) 96 | d) 99 |

7) 999,991,983,975,967,___?

| a) 989 | b) 959 |
| c) 959 | d) 990 |

8) 1111,1102,1093,1084,____?

| a) 1076 | b) 1077 |
| c) 1088 | d) 1075 |

9) 1,5,2,10,3,15,4,20,5,__?

| a) 30 | b) 25 |
| c) 35 | d) 40 |

10) -1,7, -3,14, -5,21, -7, ___?

| a) 28 | b) 35 |
| c) 21 | d) 42 |

11) 9, -1,18, -2, 27, -3, ___?

| a) 36 | b) 46 |

| c) 29 | d) 45 |

12) 5, -20,8, -60,11, -180,14, _____?

| a) 510 | b) 520 |
| c) 560 | d) 540 |

13) 9999, -1102,999, -1093,99, ___,9.

| a) 1085 | b) 1086 |
| c) 1084 | d) 1089 |

14) 1000,200,40,8, ___?

| a) 1.5 | b) 1.6 |
| c) 1.7 | d) 1.8 |

15) 9999,8000,9998,2000,9997,500, 9996, __?

| a) 125 | b) 126 |
| c) 128 | d) 129 |

16) 57.14,1111,8.16,111,1.16,11, __?

| a) 0.90 | b) 0.80 |

| c) 0.19 | d) 0.16 |

17) -111, -185, -431, __?

| a) -1293 | b) -1296 |
| c) -1950 | d) -1298 |

18) 777,97.125,12.140, ___?

| a) 1.8 | b) 1.9 |
| c) 1.2 | d) 1.5 |

19) 4,9,16,25,36,49,64, ___?

| a) 81 | b) 82 |
| c) 100 | d) 49 |

20) 8,27,64,125,216, ___?

| a) 81 | b) 343 |
| c) 345 | d) 350 |

21) 6.25,12.25,20.25,30.25, ___?

| a) 42.25 | b) 45.25 |
| c) 43.26 | d) 56.25 |

22) -64, -125, -216, -343, ____ ?

a) -513	b) 512
c) -525	d) -512

23) 1000, 1331, 1728, 2197, ____ ?

a) 2755	b) 2745
c) 2744	d) 3070

24) 0, 1, 2, 3, 5, 8, 13, 21, ____ ?

a) 34	b) 35
c) 36	d) 37

25) 0, 2.5, 2.5, 5, 7.5, 12.5, ___ ?

a) 19.5	b) 21.53
c) 18.50	d) 20.00

26) 0, 10, 11, 21, 32, 53, 85, ___ ?

a) 130	b) 138
c) 140	d) 133

27) 5.5,6,6.5,7,7.5,___?

a) 9	b) 8.5
c) 8	d) 9.5

28) 12.33,24.66,37,49.33,___?

a) 61.66	b) 62.00
c) 64.00	d) 55.50

Objective Type Number Series Answer

Answers given below in bold Letter

1) 5,8,11,14,17,20,23

5 + 3 = 8

8 + 3 = 11

$11 + 3 = 14$

$14 + 3 = 17$

$17 + 3 = 20$

$20 + 3 = 23$

$23 + 3 = \mathbf{26}$

2) 10,15,20,25,30,__,40,45.

$10 + 5 = 15$

$15 + 5 = 20$

$20 + 5 = 25$

$25 + 5 = 30$

$30 + 5 = \mathbf{35}$

$35 + 5 = 40$

$40 + 5 = 45$

3) 123,130,137,144,151,__?

$123 + 7 = 130$

$130 + 7 = 137$

$137 + 7 = 144$

$144 + 7 = 151$

$151 + 7 = \mathbf{158}$

4) 1234,1238,1242,1246,1250, __?

1234 + 4 =1238

1238 + 4 =1242

1242 + 4 =1246

1246 + 4 =1250

1250 + 4 =**1254**

5) 90,84,76,72,66, __?

90 – 6 = 84

84 – 6 = 76

76 – 6 = 72

72 – 6 = 66

66 – 6 = **60**

6) 100,98,96, ___,92,90,88.

100 – 2 = 98

98 – 2 = 96

96 – 2 = **94**

94 – 2 =92

92 – 2 =90

90 – 2 =88

7) 999, 991, 983, 975, 967, ___?

999 – 8 = 991

991 – 8 = 983

983 – 8 = 975

975 – 8 = 967

967 – 8 = **959**

8) 1111, 1102, 1093, 1084, ___?

1111 – 9 = 1102

1102 – 9 = 1093

1093 – 9 = 1084

1084 – 9 = **1075**

9) 1, 5, 2, 10, 3, 15, 4, 20, 5, __?

Solution: Divide the number in two series as given below:

Series A: 1, 2, 3, 4, 5

Series B: 5, 10, 15, 20, ……

5 x 1 = 5
5 x 2 = 10
5 x 3 = 15
5 x 4 = 20
5 x 5 = **25**

10) -1,7, -3,14, -5,21, -7, ___?

Series A: -1,-3,-5,-7

Series B: 7,14,21, ….

7x1 = 7
7x2 = 14
7x3 = 21
7x4 = **28**

11) 9, -1,18, -2, 27, -3, ___?

Series A: 9,18, 27……

Series B: -1, -2, -3

9 x 1 = 9
9 x 2 = 18

9 x 3 = 27
9 x 4 = **36**

12) 5, -20,8, -60,11, -180,14, _____?

Series A: 5,8,11,14
Series B: -20, -60, -180, ……

-20 x 1 = -20
-20 x 3 = -60
-60 x 3 = -180
-180 x 3 = **-540**

13) 9999, -1102,999, -1093,99, ___,9.

Series A: 9999,999,99,9
Series B: -1102, -1093, ……

-1102 + 9 = - 1093

$-1093 + 9 = \mathbf{-1084}$

14) 1000, 200, 40, 8, ___ ?

$1000 / 5 = 200$
$200 / 5 = 40$
$40 / 5 = 8$
$8 \ / 5 = \mathbf{1.6}$

15) 9999, 8000, 9998, 2000, 9997, 500, 9996, ___ ?

Series A: 9999, 9998, 9997, 9996
Series B: 8000, 2000, 500, …….

$8000 / 4 = 2000$
$2000 / 4 = 500$
$500 \ / 4 = \mathbf{125}$

16) 57.14,1111,8.16,111,1.16,11,__?

Series A: 57.14,8.16,1.16, ……
Series B: 1111,111,11

$57.14 / 7 = 8.16$
$8.16 / 7 = 1.16$
$1.16\ 7 = \mathbf{0.16}$

17) -111, -185, -431, __?
$-111 / 3 = (-37) \times 3 = -111$
$-111 / 3 = (-37) \times 5 = -185$
$-185 / 3 = (-61.66) \times 7 = -431$
$-431 / 3 = (143.66) \times 9 = \mathbf{-1293}$

18) 777,97.125,12.140,___?

777 / 8 = 97.125
97 / 8 = 12.140
   12 / 8 = **1.5**

19) 4,9,16,25,36,49,64,___?

2   x 2 = 4
3   x 3 = 9
4   x 4 =16
5   x 5 =25
6   x6 =36
7   x7 = 49
8   x8 =64
9   x9=**81**

20) 8,27,64,125,216,___?

2 x 2 x 2 = 8
3 x 3 x 3 = 27
4 x 4 x 4 = 64
5 x 5 x 5 = 125
6 x 6 x 6 = 216
6 x 7 x 7 = **343**

21) 6.25, 12.25, 20.25, 30.25, ___?

2.5 x 2.5 = 6.25
3.5 x 3.5 = 12.25
4.5 x 4.5 = 20.25
5.5 x 5.5 = 30.25
6.5 x 6.5 = **42.25**

22) -64, -125, -216, -343, ____?

- 4 x -4 x -4 = -64
- 5 x -5 x -5 = -125

- 6 x -6 x -6 = -216
- 7 x -7 x -7 = -343
- 8 x -8 x -8 = **-512**

23) 1000,1331,1728,2197,____?

10 x 10 x 10 = 1000
10 x 11 x 11 = 1331
11 x 12 x 12 = 1728
12 x 13 x 13 = 2197
13 x 14 x 14 = **2744**

24) 0,1,2,3,5,8,13,21,____?

0 + 1 = 1
1 + 2 = 3
3 + 2 = 5
5 + 3 = 8
8 + 5 = 13

13 + 8 = 21
21 + 13 = **34**

    25) 0,2.5,2.5,5,7.5,12.5, ___?

0 + 2.5 = 2.5
2.5 + 2.5 = 5
5 + 2.5 = 7.5
7.5 + 5 = 12.5
12.5 + 7.5 = **20**

    26) 0,10,11,21,32,53,85, ___?
0 + 10 = 10
10 + 11 = 21
21 + 11 = 32
31 + 21 = 53
53 + 32 = 85
  85   53 = **138**

27) 5.5,6,6.5,7,7.5, ___?

11 / 2 = 5.5
12 / 2 = 6
13 / 2 = 6.5
14 / 2 = 7
14 / 2 = 7.5
15 / 2 = **8**
16

28) 12.33,24.66,37,49.33, ___?

111 / 9 = 12.33
222 / 9 = 24.66
333 / 9 = 37
444 / 9 = 49.33
555 / 9 = **61.66**

WRONG NUMBER SERIES

Find out the wrong number in Series which is not suitable in the series.

1) 10,15,20,22,30,35,40

Ans: Solution as given below:

10 + 5 = 15
15 + 5 = 20
20 + 5 = **25**
25 + 5 = 30
30 + 5 = 35

35 + 5 = 40

The Wrong Number is 22 instead of 25.

2) 4,9,16,25,36,48

Ans: Solution as given below:

2 x 2 = 4
3 x 3 = 9
4 x 4 = 16
5 x 5 = 25
6 x 6 = 36
7 x 7 =**49**

The Wrong Number is 48 instead of 48.

3) 98,95,92,90,86.

Ans: Solution as given below:

98 − 3 = 95
95 − 3 = 92

92 − 3 = **89**

89 − 3 = 86

The Wrong Number is 90 instead of 89.

4) 988,997,1006,1012,1024

Ans: Solution as given below:

988 + 9 = 997

997 + 9 = 1006

1006 + 9 = **1015**

1015 + 9 = 1024

The Wrong Number is 1012 instead of 1015.

5) 1111,222.22,44.44,10.88,1.77

Ans: Solution as given below:

1111 / 5 = 222.22

222.22 / 5 = 44.44

44.44 / 5 = **8.88**

8.88 / 5 = 1.77

The Wrong Number is 10.88 instead of 8.88

6) 11,24,44,90,180

Ans: **Solution as given below:**

11 x 2 => 22 + 2 = 24
24 x 2 => 42 + 2 = 44
44 x 2 => 88 + 2 = 90
90 x 2 => 180 + 2 = **182**

The Wrong Number is 180 instead of 182.

7) 9,76,670,6106

Ans: **Solution as given below:**

9 x 9 => 81 – 5 = 76
76 x 9 => 684 -5 = **679**
679 x 9 => 6111 – 5 = 6106

The Wrong Number is 670 instead of 679.

8) 9,30,65,126

Ans: **Solution as given below:**

2 x 2 x 2 => 8 + 1 = 9
3 x 3 x 3 => 27 + 1 = **28**
4 x 4 x 4 =>64 + 1 = 65
5 x 5 x 5 => 125 + 1 = 126

The Wrong Number is 30 instead of 28.

9) 888,894,900,906,910

Ans: **Solution as given below:**

888 + 8 => 896 – 2 = 894
894 + 8 =>902 – 2 = 900
900 + 8 =>908 – 2 = 906
906 + 8 =>914 – 2 = **912**

The Wrong Number is 910 instead of 912.

SECTION 2

PERCENTAGE

"All Birds find shelter during rain. But Eagle avoids rain by flying above clouds."

- Dr APJ Abdul Kalam (Avul Pakir Jainulabdeen Abdul Kalam)

PERCENTAGE BASIC

How to calculate percentage in Examination?

Examples:

1) 10 % of 500 =?

$$\frac{500 \times 10}{100}$$

$$= 10$$

After calculation you Answer is 10.

2) 40 % of 1200 =?

$$\frac{40 \times 1200}{100}$$

$$= 480$$

After calculation your Answer is 480.

3) 80 % of 1259 =?

$$\frac{1258 \times 80}{100}$$

=1006.40

After calculation you Answer is 1006.40

4) 90 % of 1578=?

$$\frac{1578 \times 90}{100}$$

= 1420.20

After calculation you Answer is 1420.20

5) 70 % of 12.45=?

$$\frac{12.45 \times 71}{100} = 8.83$$

After calculation you Answer is 8.83

6) 5% of 1456.25=?

$$\frac{1456.25 \times 5}{100}$$

$$= 72.8$$

After calculation your Answer is 72.8

7) 7% of 12.456=?

$$\frac{12.456 \times 7}{100}$$

$$= 0.871$$

After calculation your Answer is 0.871

8) 3% of 999=?

$$\frac{999 \times 3}{100}$$

$$= 29.97$$

After calculation your Answer is 29.97

9) 8% of 7777=?

$$\frac{7777 \times 8}{100}$$

$$= 622.16$$

After calculation your Answer is 622.16

10) 2.5% of 888=?

$$\frac{888 \times 2.5}{100}$$

$$= 22.2$$

After calculation your Answer is 22.2

11) 6.5% of 9985=?

$$\frac{9985 \times 6.5}{100}$$

= 649.02

After calculation your Answer is 649.02

12) 3.5% of 4587=?

$$\frac{3.5 \times 4587}{100}$$

= 160.54

After calculation your Answer is 160.54

13) **Tom spends 30% of his monthly salary on house rent, 20% on food, 5% on transportation, 8% of Electricity Bill and 10% on other household expenses. He saves the remaining amount of Rs. 5000 at the end of the month. What is his monthly salary?**

Ans:

The Monthly Salary of Tom is 100 %

Total Expenses of Month = 30 % + 20 % + 5 % + 8 % +10 %

= 73 %

Monthly Saving = 100% - 73% = 27 %

Monthly Salary = 5000 x 100 / 27 = **18518.51**

After calculation your Answer is = 18518.51

14) The price of a car is Rs. 8,00,000. It was insured for 80% of its price. The car got completely damaged and the insurance company paid only 70% of the insured amount. What is the price of the difference between the price of the car and the amount of insurance received?

Ans:

The Value of Car is Rs 80000 & Insurance amount received 70 % of which is actual value was 80%.

⇨ 70 % of 80 %

⇨ (70 x 80 / 100) % = 56 %

The Price of Car in % is 100 %

Difference (100 – 56) % = 44 %

The difference amount Received by Insurance company is

(28 x 8,00,000) / 100 = **224000**

After calculation your Answer is 22,4000.

15) 20 % of a number is 9 less than one third of that number. Find the number.

Ans :

Let the number be x.

⇨ $(x/3) - 20\% \text{ of } x = 9$

⇨ $(x/3) - (x/5) = 9$

⇨ $X/15 = 9$

⇨ $X = 15 \times 9$

⇨ **135**

After calculation your Answer is 135.

PERCENTAGE OBJECTIVE TYPES QUESTION

1) 50 % of 300 =?

a) 145	b) 140
c) 150	d) 160

2) 90 % of 999 =?

a) 899.5	b) 899.1
c) 980	d) 956

3) 30 % of 459 =?

a) 137.7	b) 138
c) 180	d) 150

4) 80 % of 1111 =?

a) 888.8	b) 875
c) 884	d) 881.01

5) 60 % of (954 + 159) =?

a) 107.00	b) 108.90
c) 100.94	d) 105.60

6) 20 % of (999 -9) =?

a) 195	b) 198
c) 200	d) 194

7) 2.5 % of (900 / 60) =?

a) 0.250	b) 0.375
c) 0.450	d) 4.500

8) 35 % of (555 + 300 – 444) =?

a) 145	b) 143.85
c) 149.50	d) 144.50

9) 6.5 of (777 / 111 + 50 – 10) =?

a) 4.055	b) 4.560
c) 3.055	d) 3.002

10) 75 % of (88888-7777-666-55) =?

a) 60245.00	b) 60292.50
c) 60298.00	d) 60291.80

11) 20 % of (0.25 + 0.123) =?

a) 0.75	b) 0.850
c) 0.075	d) 0.749

12) 95 % of (10000 / 250 + 400) =?

a) 420	b) 415
c) 417.50	d) 418

13) Sanjay spends 10% of his monthly salary on house rent, 8% on food, 3% on transportation, 2% of Electricity Bill and 7% on other household expenses. He saves the remaining amount of Rs. 3000 at the end of the month. What is his monthly salary?

Ans:

a) 4290.20	b) 4285.71
c) 4850.50	d) 4280.90

14) Sunny buy a Car & spending on car 20 % of driver fee, 35 % of Maintenance fee, Other Expenses to driver 15 %. She is saving money from his income of Rs. 10,000 at the end of month. What is his monthly salary?

Ans:

a) 33458	b) 324546

| c) 33333.33 | d) 33458.20 |

15) 5 % of (4521 + 215 – 111)

a) 213.25	b) 251
c) 215	d) 250

PERCENTAGE OBJECTIVE TYPES QUESTION ANSWERS

1) 300 x 50 / 100 = **150**

2) 90 x 999 / 100 = **899.1**

3) 30 x 459 / 100 = **137.7**

4) 80 x 1111 = **888.8**

5) 954 + 159 = 1113

⇨ 1113 x 60 / 100 = **104.94**

6) 999 – 9 = 990

⇨ 20 x 990 / 100 = **198**

7) 900 x 60 = 15

⇨ 2.5 x 15 / 100 = **0.375**

8) 555 + 300 – 400 = 441

⇨ 35 x 441 / 100 = **143.85**

9) 777 / 111 = 7

⇨ 7 + 50 – 10 = 47

⇨ 6.5 x 47 / 100 = **3.05**

10) 88888 – 7777 – 666 – 55 = 80390

⇨ 75 x 80390 / 100 = **60292.50**

11) 0.25 + 0.123 = 0.373

⇨ 20 x 0.373 / 100 = **0.075**

12) 10000 / 250 = 40

40 + 400 = 440

⇨ 95 x 440 / 100 = **418**

13) Total Monthly Salary of Sanjay is 100 %

⇨ Total Expenses of Month is (10 + 8 + 3 + 2 + 7) %

⇨ 30%

⇨ Monthly Saving = (100 – 30) % = 70 %

⇨ 3000 x 100 / 70 = **4285.71**

14) Total Monthly Salary of Sunny is 100 %

- ⇨ Total Expenses of Car is (20 + 35 + 15) %
- ⇨ 70 %
- ⇨ Monthly Saving on Car (100 – 70) %
- ⇨ 30 %
- ⇨ 10000 x 100 / 30 = **33,333.33**

15) 4521 + 215 – 111 = 4625
- ⇨ 5 x 4625 / 100 = **213.25**

SECTION 3

PROFIT & LOSS

"Life is like A game of cards. the hand you are dealt is determinism; the way you play it is free will"

-- Pnadit Jawahar Lal Nehru

PROFIT & LOSS PRACTICE

Formula of   Profit & Loss

1) Profit = Selling Price (SP) – Cost Price(CP)

2) Loss = Cost Price(CP) − Selling Price (SP)
3) Gain % = Gain x 100 / CP
4) Loss % = Loss x 100 / CP
5) If any product sold at loss of 20 % then SP = 80 % of CP
⇨ CP = [100 / (100 + Profit %) x SP
⇨ CP = [100 / (100 - Loss %) x SP
6) Selling Price (SP) = [(100 + Profit%) x CP / 100]
7) Selling Price(SP) = [(100 − Loss%) x CP / 100]

Examples:

1) Anil Kapoor selling School Bag which is marked price of a bag is Rs.2000. A shopkeeper offers 20% discount on this bag and then again offers a 10% discount on the new price. How much will you have to pay, finally?

Ans: Solution

The successive discount is 20% and 10%
x = 20% and y = 10%
Total discount = [x + y - (xy)/100]%
Total discount = [20 + 10 - (20 x 10)/100]% = (30 - 200/100)%
= 28%
Discount = 28% of 2000 = [28/100] x 2000 = Rs.560
Selling price (SP) = marked (MP) - discount = 2000 − 560

= ☐ **1440**

2) A Patanjali Facewash marked at ☐.70 is sold for 60. Find the rate of discount.

Ans: Solution

Market Price/ Cost Price is (MP) 70 ☐

Selling Price is (SP) 60 ☐

Discount amount is = ☐ (70 – 60) = 10 ☐

Therefor Rate of Discount = Discount x 100 / MP

⇨ 10 x 100 / 70 **= 14.28** ☐

3) Salman bought a bull Dog for ☐ 80,000 and spent ☐ 3000 on its Maintenance cost. If he sells at ☐ 60,000 then what will be his loss percentage?

Ans: Solution

The Cost Price of bull Dog is = ☐ 80,000

Amount of Maintenance cost is = ☐ 3000

Total Price of Dog = (80000 + 3000) = ☐ 83000

Selling Price of Dog is = ☐ 60000

Loss = (CP – SP)

⇨ 83000 – 60000 = ☐ 23000

Loss % = Loss x 100 / CP

⇨ 23000 x 100 / 83000

⇨ 27. 71 %

4) Karina Plot sells 3/5th part of plot at a profit of 20% and remaining at a loss of 5%. If the total profit is ☐ 2,50,000 then what is the total cost price of polt?

Ans: Solution

Assume Karina be the cost price.
Therefore,
[{(3/5) x Karina x (20/100)} – {(2/5) x Karina x 5/100}]
= 2,50,000

Or Karina = ☐ **25,00,000**

5) K K Manen sold a DJ Player at a profit of 22%. Had I sold it for 28 more, 28% would have been gained. find the cost price.

Ans: Solution

Cost Price = 28 / (28-22) x 100

⇨ 28 / 6 x 100

⇨ ☐ 466.66

The Cost Price of DJ is ☐ 466.66

6) Sunny Purchase ladies Hand bag for ☐ 450 and sold for ☐ 750 Find the Profit percent.

Ans: Solution

Cost Price (CP) = ☐ 450

Selling Price (SP) = ☐ 750

Profit = (SP – CP)

⇨ 750 – 450 = ☐ 300

Profit % = Profit x 100 / CP

⇨ 300 x 100 / 450

⇨ ☐ **66.66**

7) Tom purchase a IPhone for ☐ 85,000 and he is getting 10 % profit on purchase price. Calculate the Selling price of IPhone.

Ans: Solution

Let SP = 100

⇨ SP + Profit (100 + 10) = 110

⇨ 110 % OF 85000

⇨ (110 x 85000 / 100) = ☐ **93500**

8) Shakira bought a Guitar for ☐ 8500. For how much should he sell it so as to gain 10%?

Ans: Solution

Let Coat Price (CP) = 100

Profit % = 10 %

Selling Price = (100 + 10) = 110

Cost Price is provided to us as ☐ 8500 and according to our calculation amounts to 100%. So, the equation becomes

⇨ 8500 x 110 / 100 = ☐ **9350**

Rock & Roman has partnership Business and they sold a Gym article at a loss of 8%. If the selling price had been increased by ☐ 100, there would have been a gain of 2%. What was the cost price of the Gym article?
Ans: Solution

Let CP be ☐ Y then (102 % of Y) – (92 % of Y) = 100

⇨ 10108 / 100 = ☐ **101.08**

9) Sharukh fixes the marked price of an item 30% above its cost price. The percentage of discount allowed to gain 15% is.

Ans: Solution

Let the Cost Price (CP) = 100

then Market Price(MP) = 130

required gain = 15%

so selling Price(SP) = 115

Discount price = (MP − SP)

⇨   130 − 115 = 15

Discount % = (Discount / MP) x 100

⇨   15 / 130 x 100

⇨   **11.53 %**

PROFIT & LOSS OBJECTIVE TYPE QUESTIONS

1) Raj Bawaja Sold a mobile at ☐ 15,000. if the cost of price is 12000. find out Profit?

Ans:

a) 4500	b) 3000
c) 3800	d) 2900

2) Bharat Singh sold a one pair shoes at ☐ 550. If the cost of price of shoes is ☐ 750. Find out loss?

Ans:

a) 200	b) 250
c) 350	d) 450

3) Sweety Sold Hand bag at 750. if the coat of price of bag is ☐ 880. find out Loss Percent.

Ans:

a) 15.88 %	b) 14.77 %
c) 16.55 %	d) 13.99 %

4) Afrin sold a Hand watch at 5500. If the cost of price is 5000. Find out profit percent.

Ans:

| a) 12 % | b) 18 % |
| c) 15 % | d) 10 % |

5) A Patanjali eco-friendly cracker marked at ☐.40 is sold for 25. Find the rate of discount.

Ans:

| a) 37.50 | b) 60.50 |
| c) 35.40 | d) 36.90 |

1) Aman Kary selling Eco-Friendly Tube light which is market price of a Tube light is Rs. 1500. A shopkeeper offers 15% discount on this Tube light and then again offers a 5% discount on the new price. How much will you have to pay, finally?

| a) 1215 | b) 1240 |
| c) 1250 | d) 1280 |

Vikash bought a Car for ☐ 2,80,000 and spent ☐ 3000 on its Maintenance cost. If he sells at ☐ 1,50,000 then what will be his loss percentage?

Ans:

| 45.00 % | 47.99 % |
| 48.99 % | 46.75 % |

Mathematics One Day, Page | 75

Pradeep Professional Camera sells 3/5th part of Camera at a profit of 30% and remaining at a loss of 5%. If the total profit is ☐ 1,50,000 then what is the total cost price of Camera?
Ans:

925002	937450
937500	934000

1) Ajay Shulka sold a Power cable at a profit of 20%. Had I sold it for 40 more, 40% would have been gained. find the cost price.

Ans:

a) 200	b) 400
b) 198	c) 210

1(a) Ansu Purchase Mobile bag for ☐ 7500 and sold for ☐ 8500 Find the Profit percent.
Ans:

13.00	13.33
14.55	12.99

2) Mr Anil Shukla purchase a White Car for ☐ 3,85,000 and he is getting 6 % profit on purchase price. Calculate the Selling price of Car.

Ans:

a) 4,08,100	b) 4,09,200
c) 4,08,200	d) 4,08,980

3) Vinod & Bunty bought a Movie Camera for ☐ 2,10,00. For how much should he sell it so as to gain 20%?

a) 2,51,500	b) 2,35,000
c) 2,51,998	d) 2,52,000

4) Dharmahas partnership Business and they sold a Rickshaw article at a loss of 9%. If the selling price had been increased by ☐ 200, there would have been a gain of 5%. What was the cost price of the Rickshaw article?

Ans:

a) 714.18	b) 715
c) 712	d) 740

Chanchal fixes the marked price of an item 40% above its cost price. The percentage of discount allowed to gain 15% is.

Ans:

a) 17.95	b) 17.85
c) 18.95	d) 20.85

PROFIT & LOSS OBJECTIVE TYPE QUESTIONS ANSWERS

Answer (C)
Cost of Goods Price (Cp) = 12000
Selling Price (SP) = 15000

Profit = SP – CP

⇨ 15000 – 12000 =

⇨ ☐ **3000**

1) **Answer (A)**

Cost of Price of Shoes (CP) = 750

Selling Price (SP) = 550

Loss = (CP – SP)

⇨ 750 – 550 =

⇨ ☐ 200

2) Answer (B)

Cost of Price of Bags (CP) = ☐ 880

Selling Price (SP) = ☐ 750

Loss = (CP – SP)

⇨ 880 – 750 = 130

Loss % = Loss x 100 / CP

⇨ 130 x 100 / 880 =

⇨ **14.77 %**

3) ANSWER (D)

Cost of price of Watch (CP) = 5000

Selling Price(SP) = 5500

Profit (SP – CP)

⇨ 5500 – 5000 =

⇨ 500

Profit % = Profit x 100 / CP

⇨ 500 x 100 / 5000

⇨ **10 %**

4) Answer (A)

Market Price/ Cost Price is (MP) = 40 ▢

Selling Price is (SP) = 25 ▢

Discount amount is = ▢ (40 – 25) = 15 ▢

Therefor Rate of Discount = Discount x 100 / MP

⇨   15 x 100 / 40 **= 37.50** ▢

5) Answer (A)

The successive discount is 15% and 5%
x = 15% and y = 5%
Total discount = [x + y - (xy)/100]%
Total discount = [15 + 5 - (15 x 5)/100]% =

⇨   (20 - 75/100) %

⇨   19.25 %

⇨   Round off = 19 %

Discount = 19% of 1500 = [19/100] x 1500 = Rs.285
Selling price (SP) = marked (MP) - discount = 1500 – 285

= ▢ **1215**

6) Answer (B)

The Cost Price of Car is = ₹ 2,80,000

Amount of Maintenance cost is = ₹ 3000

Total Price of Car = (2,80000 + 3000) = ₹ 2,83,000

Selling Price of Car is = ₹ 1,50,000

Loss = (CP – SP)

⇨ 2,83000 – 1,50,000 = ₹ 1,33,000

Loss % = Loss x 100 / CP

⇨ 1,33,000 x 100 / 2,83,000

⇨ **46. 99 %**

7) Answer (c)

Assume Predeep be the cost price.
Therefore,
[{(3/5) x Pradeep x (30/100)} – {(2/5) x Pradeep x 5/100}]

= 1,50,000

⇨ 8 / 50 x 1,50,000 = **9,37,500**

8) Answer (A)

Cost Price = 40 / (40-20) x 100

⇨ 40 / 20 x 100

⇨ ▢ 200.00

The Cost Price of DJ is ▢ **200.00**

9) Answer (B)

Cost Price (CP) = ▢ 7500

Selling Price (SP) = ▢ 8500

Profit = (SP – CP)

⇨ 8500 – 7500 = ▢ 1000

Profit % = Profit x 100 / CP

⇨ 1000 x 100 / 7500

▢ **13.33 %**

10) Answer (A)

Let SP = 100

⇨ SP + Profit (100 + 6) = 106

⇨ 106 % OF 3,85,000

⇨ (106 x 3,85,000 / 100) = ₹ **4,08,100**

11) Answer (D)

Let Coat Price (CP) = 100

Profit % = 20 %

Selling Price = (100 + 20) = 120

Cost Price is provided to us as ₹ 2,10,00 and according to our calculation amounts to 100%. So, the equation becomes

⇨ 2,10,000 x 120 / 100 = ₹ **2,52,000**

12) Answer (A)

Let CP be ₹ Z then (105 % of Z) – (91 % of Z) = 100

⇨ 0.14 Z = 100

⇨ Z = 100 x 100 / 14

⇨ ₹ **714.28**

13) ANSWER (B)

Let the Cost Price (CP) = 100

then Market Price(MP) = 140

required gain = 15%

so selling Price(SP) = 115

Discount price = (MP – SP)

⇨ 140 – 115 = 25

Discount % = (Discount / MP) x 100

⇨ 25 / 140 x 100

17.85 %

SECTION 3

AVERAGE

"One Individual may die for an Idea; but that idea will, after his death, incarnate itself in a thousand lives"

-- Netaji Subash Chandra Bose

AVERAGE PRACTICE

Average Formula:

$$= \frac{\text{Sum of Terms}}{\text{Number count}}$$

1) 12,15,18,21 calculate the average value as given series.

Ans: Solution

Formua = Sum of Terms / Number Count

⇨ 12 + 15 + 18 + 21 / 4
⇨ **16.5**

2) **m + n = 12 , n + p = 18, m + p = 22 What is the average of m,n and P.**

Ans : Solution

⇨ M + n = 12 --------------------(1)
⇨ N + p = 18 ---------------------(2)
⇨ M + p = 22 ---------------------(3)

If we are adding 3 equation, we get it ,

⇨ 2m + 2n + 2p = 12 + 18 + 22
⇨ 2 (m + n + p) = 52
⇨ m + n + p = 52/2

⇨ ∴ m + n + p = 26

Hance Average of m,n,p = m+n+p / 3

⇨ 26 / 3
⇨ **8.66**

3) **There are 40 employees in LLPP. If 15 new employee joined, the total expenditure increased by ☐ 1500 while the average expenditure increased by 20 ☐ . What was the initial average expenditure of employee?**

Ans: Solution

Let the Initial Expenditure = y

Total Initial Expenditure = 40 y

Final Total Expenditure = 40 y + 1500

Average Expenditure = y + 20

We Know that

Average = Sum of Terms / Number Count

⇨ Y + 20 = 40 y + 1500 / (40 + 15)

⇨ 55y + 1100 = 40y + 1500
⇨ Y = 400 /15 = **26.66**

4) The first 6 overs of a twenty – twenty cricket game, the run rate was only 6.0. What should be the run rate in the remaining 14 overs to reach the target of 210 runs?

Ans:

The runs scored in the first 6 overs = 6 × 6.0 = 36
Total runs = 210

Remaining runs to be scored = 210 - 36 = 174
Remaining overs = 14

Run rate needed = 174 / 14 = **12.42**

5) The average weight of 10 travel bags in a Dubai Mall is 5.50 kg and that of the remaining 15 travel bags is 2.50 kg. Find the average weights of all the travel bags in the Dubai Mall.

Ans:

Average weight of 10 travel bags = 5.50
Total weight of 10 travel bags = 5.50 × 10

Average weight of remaining 15 travel bags = 2.50
Total weight of remaining 15 travel bags = 2.50 × 15

Total weight of all travel bags in the Dubai Mall = (5.50 × 10) + (2.50 × 15)

Total travel bags = 10 + 15 = 25

Average weight of all the travel bags: =

(5.50 × 10) + (2.50 × 15) / 25

= 55 + 37.50 / 25

= **3.7**

6) **British library has an average of 200 visitors on Monday and 150 on other days. What is the average number of visitors per day in a month of 30 days beginning with a Monday?**

Ans:

In a month of 30 days beginning with a Monday, there will be 4 complete weeks and another two days which will be Sunday and Monday.

Hence there will be 5 Monday and 25 other days in a month of 30 days beginning with a Monday

Average visitors on Monday = 200
Total visitors of 5 Monday = 200 × 5

Average visitors on other days = 150
Total visitors of other 25 days = 150 × 25

Total visitors = (200 × 5) + (150 × 25)
Total days = 30

Average number of visitors per day =

(200 x 5) + (150 x 25) / 30

= 1000 + 3750 / 30

= **158.33**

7) **The average of eight Digital IPad numbers id 27. If two IPad number is excluded, the average becomes 15. What is the excluded number?**

Ans:

The Sum of 5 number Digital IPad = 5 x 27

The sum of 6 number Digital IPad after Excluding 2 number = 6 x 15

Excluded Number = (27 x 5) – (6 x 15)

= 135 − 90

= **45**

8) **The average of 22 numbers is 35. If two numbers, 4 and 10 are discarded, then the average of the remaining numbers is nearly:**

Ans:

The total sum of 20 Number = (22 x 15) − (4 x 10)

⇨ 330 − 40
⇨ 290

Average = Total Sum / Count Number

⇨ 290 / 20
⇨ **14.5**

The Answer will be *14.5*

SECTION 4

TIME & DISTANCE

"Manpower without unity is not a strength unless it is harmonized and united properly, then it became a spiritual power"

- Sardar Patel (Vallabhbhai Jhaverbhai Patel)

TIME SPEED & DISTANCE

Formula:

1) Distance = Speed x Time

2) Distance = Rate x Time

3) Speed = Distance / Time

4) Time = Distance / Speed

Conversation Unit:

1) 1 Km/h = 5/18 meter/Second
2) 1 meter/second = 18/5 Km/hour
3) 1 km/hour = 5/8 mile / hour
4) 1 mile /hour = 22/5 foot/second

Basic Conversation Unit List:

Time = Second , Minutes , Hours

Distance = meter & kilometer

Speed = Kilometer/ hour or meter/second

Convert kilometers per hour(km/hour) to meters per second(m/s)

x km/hr

- ⇨ x km in 1 hour
- ⇨ 1000 x meter(m) in 3600 second(s)
- ⇨ 1000x3600 meter(m) in 1 second(s)
- ⇨ 5x18 meter(m) in 1 second(s)
- ⇨ x×518 m/s

1) **Bike travels at the speed of 75 kmph(Kilometer per hours). What is distance covered by the bike in 5 minutes.**

Ans:

The Speed of Bike = 75 Kmph.

= 75 x 1000 / 60

= 1250 meter/minutes

.: Distance covered in 5 minutes = 5 x 1250 = 6250 meter

The Answer is **6250** meter.

2) **One Hoarse Covering distance at the rate of 5 kmph a man cover certain distance in 2 hr 45 min. Running at a speed of 7.5 kmph the man will cover the same distance in.**

Ans:

Distance = Speed x Time

Here time = 2 hour and 45 minutes = 11/4

Distance = 5 x 11 / 4 = 13.75 km

New Speed = 7.5 kmph

Therefor time = Distance / Speed

⇨ 13.75 / 7.5
⇨ 1.8 minutes

The answer is **1.8 minutes**.

3) **Two Car championships Riesling starting at the same time from 2 Car stations 300 km apart and going in opposite direction cross each other at a distance of 160 km from one of the stations. What is the ratio of their speeds?**

Ans:

In same time Car Riesling, they cover 300 km and 140 km respectively.

For the same time speed and distance is inversely proportional.

So ratio of their speed = 300 : 140

⇨ 30 : 14
⇨ 15 : 7

The answer is **15: 7 Ratio**.

4) **The Ratio between two Boats is 2:5. if the second boat run 150 km in 6 hours, then what is speed of first boat.**

Ans:

Let the speed of two boats is = 2x and 5x km/h

Then,

⇨ 5x = 150/ 6
⇨ 5x = 25
⇨ X = 25 / 5
⇨ X = 5

The answer is **5 km/h**

5) **The Prince drive a car 120 meter in 50 second. What is the speed of car?**

Ans:

Speed = Distance / Time

= 120 / 50 m/sec

= 12/ 5 x 18 / 5 km/hour

= 216 / 25 km/hour

= 8.64 km/hour

6) **Sudhir riding Ducati Testa bike is 110 meter approx. 20 meter/second . How much time it will take?**

Ans:

Time = Distance / Speed

= 120 / 20

= 6 m/s

= 6 x 18 / 5

= 108 / 5
= 21.06

The Answer **is 21.**

7) **One Train travelling on first 110 km at 40 km/hr and the next 70 km at 60 km/hr. Find the average speed for first 180 km of train.**

Ans:

Time = Distance / Speed

So the time taken = (110/40 + 70/60)

⇨   2150 / 4 hours

Time Taken 180 Km/hours
⇨   180 * 4 / 2150 = **0.059**

8) **The Speed of train is 30 meter per second. It can cross a pole within 8 second what is the length of train.**

Ans:

The Length of Train = 30 x 8 = 240 meter

The answer is **240 meter.**

9) **Sachin & Dhananjay running at 5 km/h takes one hour to cover a distance. If the speed is reduced by 3 km/hour then in how much time it will cover the distance ?**

Ans:

Reduce Speed = 5 – 3

= 2 Km/h

The new speed = 5 / 2 x 60

= 300/ 2

= 150 minutes

The Answer is **15 minutes**.

10) State championship race run by Student, Diwaker beats Anoop by 100 meters. Anoop beats Akhilesh by 100 meters. By how much meters does Diwaker beat Akhilesh in the same race?

Ans:

⇨ While Diwaker covers 1000 meters, Anoop can cover 900 meters
⇨ While Anoop covers 1000 meters, C can cover 900 meters

Lets assume that all three of them are running same race.

So when Anoop runs 900 meters, Akhilesh can run 900 × 9/10 = 810

So Diwaker can beat Akhilesh by 190 meters.

The Answer is **190 meters.**

11) **The LLP College announced Yearly skating competition for one kilometer. Rajnesh beats Salman by 60 meter or 6 seconds. Find the Rajnesh time over the game.**

Ans:

Rajnesh beats Salman by 60 meter or 6 second, so Salman take 6 second to cover the distance in 60 meter.

Hence, speed of Salman = 60 / 6 = 10 m/s

Thus, the time spend by Salman = 1000/ 6 = 166.66 second

Time taken by Rajnesh = 167 – 6 = 161 second

The Answer is **161 second.**

12) A Women travels 800 km at 50kmph. He then travels another 500km at 30kmph. What is the average speed?

Ans:

Time taken for 800 km = 800 / 60 = 13.33 hours

Time taken for 500 km = 500 / 30 = 16.66 hours

Since the time is same in both cases, use the above formula to find out the average speed.

S1 = 50

S2 = 30

Speed = (50+30)/2 = 40kmph

The Answer is **40kmph.**

SECTION 5:

BASIC ALGEBRA

"The greatness of humanity is not in being human, but in being humane."

Mahatma Gandhi

FORMULA:

1) The Area of rectangle (A) : W x H

Where is W = Width & H = Height

2) The Area of Circle = πr^2

Where is π = 3.1415 & R = Radius

3) $(a + b)^2 = a^2 + b^2 + 2ab$
4) $(a-b)^2 = a^2 + b^2 - 2ab$
5) $a^2 - b^2 = (a + b)(a - b)$
6) $(a + b + c)^2 = a^2 + b^2 + c^2 + 2ab + 2bc + 2ca$

7) $(a + b)^3 = a^3 + 3a^2b + 3ab^2 + b^3$
8) $(a - b)^3 = a^3 - 3a^2b + 3ab^2 - b^3$
9) $a^3 - b^3 = (a - b)(a^2 + ab + b^2)$
10) $a^3 + b^3 = (a + b)(a^2 - ab + b^2)$

Examples:

Simplify the equation:

1) $z^2 + 9z + 20 = 0$

Ans:

⇨ $Z^2 + 4Z + 5Z + 20$
⇨ $Z(z+4)\; 5(z + 4)$
⇨ $(z+4)(z+5)$

2) The ratio of age of man and women was 3:2 four year ago. After 4 years the ratio will become 6:4. What is the age of women.

Ans: Before 4-year age of man and women was 3x and 2x.

After 4 years, 8 year will be added to ages of bath genders.

So, According for question –

$3x + 8 : 2x + 8 = 6 : 4$

⇨ $3x + 8 / 2x + 8 = 6/4$
⇨ $4(3x + 8) = 6(2x + 8)$
⇨ $X = 48 - 32$
⇨ $X = 6$

Hence present age is $x + 4$ or $6 + 5 = $ **11**

3) **The sum of 2 number is 120. If one – 3rd of first number exceed one seventh of second number by 10. find smaller number.**

Ans: Let the number be y and $(120 - y)$ then

According to questions

⇨ $Y/3 - (120 - y)\,7 = 10$
⇨ $7y - 3(120 - y) = 10$
⇨ $7y - 360 - 3y = 10$

⇨ 4y = 360 +10
⇨ 4y = 370
⇨ Y = 370/4

So, y = 92.50 or 92 Approx. Hence, The smaller number will be **92**

4) **The present age of Swati is 2 times the age of Sunny. The present age of Sunny is 22 years less then age. what is the age of Swati?**

Ans: Lets present age of Sunny is Z year.

Now, The age of Swati is 2 year

The difference between the age of Swati and Sunny = 22 year

So, 2z – z = 22

⇨ Z = 22

So, present age of Sunny is 22 year

The age of Swati = 2Z

⇨ 2 x 22
⇨ 44 year

The Age of Swati = **44 year**

5) Solve the Equation and find the value of X as given below : x2 +25

Ans: $(x)^2 - (5)^2 = 0$

⇨ $(x-5)(x+5) = 0$

Then x = 5 or x = -5

6) Find the value of area for Industrial Cylinder whose circumference is 50 cm and height is 15 cm.

Ans: Circumference = 2πr = 50 cm & Height = 15 cm

The area of cylinder = 2πrh

So, 50 x 15 = **750 cm**

Section 6:

Take up one idea. Make that one idea your life – think of it, dream of it, and live on that idea. Let the brain, muscles, nerves, every part of your body, be full of that idea, and just leave every other idea alone. This is the way to success.

- Swami Vivekananda

Time and Work Series

Formulas:

1) Work Done = Rate x Time

2) Constant Formula of MDH = M * D * H / W

Where,

M = Number of Man worked

D = Number of Day

H = Number of hour taking for works

W = Amount of Work taking

3) Percentage (%) of work done meaning tricks:

½ = 50 %	1/3 = 33.33 %	1/4 = 25 %	1/5 = 20 %
1/6 = 16.66 %	1/7 = 14.28 %	1/8 = 12.5 %	1/9 = 11.11 %

Examples:

1) Salim twice is efficient as Salma Begam and Salim takes 25 days to do a job, then in how many days Salma Begam can finish the same job?

Ans: The Ratio of efficiency of Salim: Salma Begam = 2:1

So, Ratio of required days of Salim: Salma Begam = 2/1 : 1/1 = 1:2

Since, Salim take 25 days, so Salma Begam will take (2x25) = **50** days to finish the same job.

2) Jay and Viru together can complete a piece of work in 10 days. If Jay alone can complete the same work in 30 days, in how many days can Viru alone complete that work?

Ans: As Pre Question Jay and Viru (Jay + Viru) working together in 1 days work = 1/10

Jay 1 days work = 1/30

So, Viru one day work = 1/10 – 1/30 = 1/15

Hence, Viru can complete work in **15 days**

3) **Vijay does a work in 6 days and Rajesh does the same work in 4 days. In how many days they together will do the same work?**

Ans: Vijay one day works = 1/6

Rajesh One day works = 1/4

If they working together = (1/6 + ¼) = 10/24

So, **5/12** Approx. 5 day 12 hour

4) Ashish, Narayan and Jayanto can complete a piece of work in 6, 3 and 4 days respectively. If they Working together, how much time it will complete the same work.

Ans: (Ashish + Narayan + Jayanto) 's one day work =

(1/6 + 1/3 + 1/4) = 54/72

So, if they complete the same work together = **72/54 days**

5) Vikash can lay Metro track between two given stations in 16 days and Anil can do the same job in 12 days. With the help of Niraj, they did the job in 4 days only. Then, how much time it will to complete the job by Niraj if they work alone.

Ans: (Vikash + Anil +Niraj) one day work = ¼

Vikash One day work = 1/16

Anil One day work = 1/12

So, one day work of three = ¼ - (1/16 +1/12)

> ⇨ (1/4 − 7/48)
> ⇨ 5/48

So, Niraj can do the work in **48/9 days.**

6) **6 Girls and 3 boys working together can do four times as much work as a girls and a boy. Working capacity of girls and boy is in the ratio.**

Ans:

Let 1 girls 1 day work = x
1 boy 1 day work = y

then 6x + 3y = 4(x+y)

> ⇨ 6x + 3y = 4x + 4y
> ⇨ 6x − 4x = 4y − 3y
> ⇨ 2x = y
> ⇨ x/y = **1/2**

So, Ratio will be **1: 2**

7) Any Army Camp if 25 soldier can do piece of work in 36 days working 10 hours a day, then how many soldier are required to complete the working 6 hours a day in 20 days.

Ans: M1 x D1 x H1 = M2 x D2 x H2

25 x 36 x 10 = M2 x 20 x 6

M2 = **75 Person**

The 75 Person required for complete the task.

8) A Event Group of women can complete the job in 120 days. if there were 4 more such women then the work could be finished in 12 days less. What was the actual strength(Power) of women's?

Ans: M 1 x D 1 = M2 x D2

X x 120 = (x + 4) x 108

So, x = **36**

The Answer will be 36 more women required.

SECTION 7:

"I'm not a handsome guy, but I can give my hand to someone who needs help. Beauty is in the heart, not in the face."

- **Dr. APJ Abdul Kalam**

Ratio & Proration Series

1) **Sanjay earn money from LIC of ☐ 10,000 per month and Anil earn only ☐ 5000 per month. Find out the ratio of between them.**

Ans: Required Ratio = 10000 / 5000 = **1:2**

2) **The MMS company out of 150 person working in an office for mobile operation.25 are men and remain are young women. Find out the ratio of number of young women to the number of men.**

Ans: The total person working = 150

Women working = 150 − 50 = 100

⇨ Ratio = 100/25
⇨ ¼

So, Ratio will be **1:4**

3) **Simplify the ratio as given below: 1/7 : 1/3**

Ans: 1/7 : 1/3 = 21/7 : 21/3 = **3:7**

4) The women selling 22 orange is being divided among two men in the ratio of 1/8 : 1/3. How much orange does each get?

Ans: 1/8 : 1/3 (The LCM of 8 & 3 = 24)

Now, 1/8 x 24 : 1/3 x 24 = 3/8

So,
The ratio of men1 will be 3/11 x 22 = **6**
The ratio of man 2 will be 8/11 x 22 = **16**

5) Find the value of ratio as given below A:B = 4:5, B:C = 3:2, C:D = 6:7.

Ans:
A:B = 4:5
B:C = 3:2
C:D = 6:7
A:B:C:D = (4 X 3 X 6) : (5 X 3 X 6) : (5 X 2 X 3) : (5 X 2 X 7)

So, A:B:C:D = **72 : 75 : 30 : 70**

6) Paro Selling milk, there are two type of mixture of milk and water. In the first mixture, out of 10 liters of mixture, 4 liter of milk only and in this second mixture 5 liter is milk and 10 liters is water. Which one mixture is batter in term of energy milk.

Ans: First Mixture = 4/10

Second mixture = 5/14 (Mixture =milk + water so, (5 + 10 = 14)

⇨ 4/10 x 70 : 5/14 x 70
⇨ 28: 25

7) The Monthly income of Tanya and Tina are in the ratio of 4:5 and their saving are in the ratio of 1:2. If the expenditure of each will be ☐ 15000, then the find out the monthly Income.

Ans: Income = Expenditure + Saving

Tanya Income 4x = y + 15000

Tina Income 5x = 2y + 15000

Therefor, 4x − y = 15000 and 5x − 2y = 15000

⇨ 4x − y = 5x − 2y
⇨ 2y − y = 5x − 4x
⇨ Y = x

Then, The Income of Tanya = 4 x 15000 = **60,000**
The Income of Tina = 5 x 15000 = **75,000**

8) **Khesari and Karenjit Kaur doing milk mixture business. They are doing mixture milk & water. The concentration of milk in each of the containers is 40 % 70% respectively. What is the ratio of water in both the containers respectively?**

Ans: The Percentage of milk and water as given below:

Milk 40 % and 70 %
Water 60 % and 30%

Therefor, require ratio = 60 / 30 , 1:2

9) Find out the ratio of Ranjit and his friends Santosh is 3:5. The difference of their ages is 8 years. What will be ratio of their ages after 2 years?

Ans:

Lets age of Ranjit is 3x and Santosh age is 5x after 2 year age will be (3x +2) and (5x +2) respectively.

Then,

⇨ 3x + 5x = 8
⇨ 8x = 8
⇨ X = 1

Therefor, the ratio of their age 2 year after

3x + 2 / 5x + 2 = **5/7**

10) Sashi traveling by ship from Dubai to Mumbai the total distance of 250 mils in 8 hour. Partial they

travel by car at 30 miles/hr. find out distance of traveler of car.

Ans: x / 30 + 250 + x /35 = 8

X = **180 miles/hour.**

SECTION 8:

"Do not be proud of wealth, people, relations, and friends, or youth. All these are snatched by time in the blink of an eye. Giving up this illusory world, know and attain the Supreme."

- Adi Guru Sankaracharya

Simple & Compound Interest

Formula:

1) Simple Interest (SI) = P x R x T / 100

Where,

P = Principle
R = Rate of Interest
T = Time

2) Amount(A) = P + prt/100 = P (1+ rt/100)

3) Compound Interest (CI) = A − P

Where,

A = Amount
P = Principle Amount

4) A = P(1 + r/100) t

5) When we calculate ½ rate of interest of CI

A = P (1+ r/2/100)^{2t}

> **Note:** We can calculate value as same on changing formula number 5 for quarter 4ᵗʰ T and yearly as per year specified.

6) The Difference between Compound Interest & Simple Interest between 2-year formula.

$= P(r/100)^2$

7) The Difference between Compound Interest & Simple Interest between 3-year formula.

$= P(r/100)^2 (r/100 + 3)$

Examples:

1) Aman Khary paying rent of 6000 to Amish Tripathi for 2 years and ☐ 4000 to Alok Tripathi for 4 years on simple interest at the same rate of interest and received ☐ 2200 in all from both of them as interest. Calculate the rate of interest per annum.

Ans: Lets R % is simple interest of Rate

Where, SI = P x R x T /100

  ⇨ 6000 x R x 2 / 100 + 4000 x R x 4 / 100 = 2200
  ⇨ 120 R + 160 R = 2200
  ⇨ 280 R = 2200

So, R = 2200/280

R = 7.8 %

2) If a sum of money doubles itself in 6 years at simple interest, then what is rate of interest of per annum.

Ans: Let sum = y then Simple Interest = y

As per Question,

Rate = (100 x y) / (y * 6)

- ⇨ 100 y = 6 y
- ⇨ Y = **16.66 %**
3) **Sudhir Prihar took a loan for 5 years at the rate of 12.00% per annum on Simple Interest, If the total interest paid was Rs. 50,000 Calculate the principal Amount.**

Ans: SI = P x R x T /100

- ⇨ P = SI x 100 / R x T
- ⇨ P = 50,000 x 100 / 12 x 5
- ⇨ P = 30,00,000

So, Principle Amount will be ☐ 30,00,000.

4) **What is the difference between simple interest received from two different sources on ☐ 2000 for 3 years is Rs.12.50. Calculate the difference between their rates of interest.**

Ans: Two year Interest = 2000 x 3 = 6000

⇨ $6000(R_1 - R_2) = 1250$
⇨ $R_1 - R_2 = 6000 / 1250$
⇨ $R_1 - R_2 = 4.8$

So, The Difference Rate of Interest will be = **4.8 %**

5) **Hemant & MF Hussain sum was put at simple interest at a certain rate for 4 years. Had it been put at 3% higher rate, it would have fetched ₹ 500 more. Find the sum of Amount.**

Ans: SI = P x R X T / 100

⇨ $P \times (R + 3) 4 / 100 - P \times R \times 4 /100 = 500$
⇨ $4P \times (R + 3) - 4PR = 500 \times 100$
⇨ $4PR + 12P - 4PR = 50000$
⇨ $12 P = 50000$
⇨ $P = 50000 / 12$

So, P = 4166.66

The Sum of amount will be ₹ **4167 /-**

6) **Bhawani Singh taking money from friends of sum of ▢ 15,000 amounts to ▢ 25,000 in 2 years at the rate of simple interest. Calculate the rate of interest?**

Ans: SI = (25000 – 15000) = 10,000

Rate = (100 x 10000) / 15000 x 2

⇨ **33.33 %**

The Rate of Interest will be **33.33 %**

7) **A certain sum of money becomes four times of itself in 30 years at simple interest. In how many years does it become double of itself at the same rate of simple interest?**

Ans: t4 = n2 – 1 / n1- 1 x T1

⇨ 2- 1 / 4 -1 x 30
⇨ 1/3 x30
⇨ 10 year

8) Calculate the time, In what time will the simple interest be 4/2 of the principal at 10 percent per annum?

Ans: Let SI = 4 y

Principle = 2 y
Rate = 9 %
T = ?

4 y = 2y x 10 x T /100

⇨ T = **20 Year**

9) Dhananjay taking loan for compound interest on ▢ 30000 at 7% per annum for a certain time is ▢ 4347. The time is?

Ans: A = CI + Principle

⇨ 30000 + 4347
⇨ 34347 ▢

$A = P(1 + R/t) 100$

$34347 = 30000 (1 + 7/100) t$

$34347/30000 = (107/100)t$

$(107/100)^2 = (107/100)^2$

T = 2 year

10) Ashish buy a bike of ☐ 10,000 at 8% rate of interest p.a. compounded annually which is paid back in 3 equal annual instalment. What is the amount of each installment?

Ans: $10,000 (1.08)^3 = y [1 + (1.08) + (1.08)^2]$

⇨ Y = 3880.35

So, The Installment will ☐ **3880 /-**

TRICKS OF NUMBER

The Number trick used for first calculation of mathematics during the examination. You should have to remember few numbers as given below:

a) Square numbers:

$2^2 = 4$	$3^2 = 9$	$4^2 = 16$	$5^2 = 25$	$6^2 = 36$
$7^2 = 49$	$8^2 = 64$	$9^2 = 81$	$10^2 = 100$	$11^2 = 121$
$12^2 = 144$	$13^2 = 169$	$14^2 = 196$	$15^2 = 225$	$16^2 = 256$
$17^2 = 289$	$18^2 = 324$	$19^2 = 361$	$20^2 = 400$	$21^2 = 441$
$22^2 = 484$	$23^2 = 529$	$24^2 = 576$	$25^2 = 625$	$26^2 = 676$
$27^2 = 729$	$28^2 = 784$	$29^2 = 841$	$30^2 = 900$	

b) Cube Numbers:

$2^3 = 8$	$3^3 = 27$	$4^3 = 64$
$5^3 = 125$	$6^3 = 216$	$7^3 = 343$
$8^3 = 512$	$9^3 = 729$	$10^3 = 1000$
$11^3 = 1331$	$16^3 = 4096$	$21^3 = 9261$
$12^3 = 1728$	$17^3 = 4913$	$22^3 = 10648$
$13^3 = 2197$	$18^3 = 5832$	$23^3 = 12167$
$14^3 = 2744$	$19^3 = 6859$	$24^3 = 13824$
$15^3 = 3375$	$20^3 = 8000$	$25^3 = 15625$

Formulas Collection # 1

Formula of Profit & Loss

1) Profit = Selling Price (SP) – Cost Price(CP)
2) Loss = Cost Price(CP) – Selling Price (SP)
4) Gain % = Gain x 100 / CP

5) Loss % = Loss x 100 / CP
6) If any product sold at loss of 20 % then SP = 80 % of CP
⇨ CP = [100 / (100 + Profit %) x SP
⇨ CP = [100 / (100 - Loss %) x SP
7) Selling Price (SP) = [(100 + Profit%) x CP / 100]
   Selling Price(SP) = [(100 – Loss%) x CP / 100]

Average Formula

$$= \frac{\text{Sum of Terms}}{\text{Number count}}$$

Time Speed & Distance

Formula:

1) Distance = Speed x Time

2) Distance = Rate x Time

3) Speed = Distance / Time

4) Time = Distance / Speed

Conversation Unit:

1) 1 Km/h = 5/18 meter/Second
2) 1 meter/second = 18/5 Km/hour
3) 1 km/hour = 5/8 mile / hour
4) 1 mile /hour = 22/5 foot/second

Basic Conversation Unit List:

Time = Second, Minutes, Hours

Distance = meter & kilometer

Speed = Kilometer/ hour or meter/second

Formula of Basic Algebra

1) The Area of rectangle (A) : W x H

Where is W = Width & H = Height

2) The Area of Circle = πr^2

Where is π = 3.1415 & R = Radius

3) $(a + b)^2 = a^2 + b^2 + 2ab$
4) $(a-b)^2 = a^2 + b^2 - 2ab$
5) $a^2 - b^2 = (a + b)(a - b)$
6) $(a + b + c)^2 = a^2 + b^2 + c^2 + 2ab + 2bc + 2ca$
7) $(a + b)^3 = a^3 + 3a^2b + 3ab^2 + b^3$
8) $(a - b)^3 = a^3 - 3a^2b + 3ab^2 - b^3$
9) $a^3 - b^3 = (a - b)(a^2 + ab + b^2)$
10) $a^3 + b^3 = (a + b)(a^2 - ab + b^2)$

Time and work series

Formulas:

1) **Work Done = Rate x Time**

2) **Constant Formula of MDH = M * D * H / W**

Where,

M = Number of Man worked

D = Number of Day

H = Number of hour taking for works

W = Amount of Work taking

3) **Percentage (%) of work done meaning tricks:**

½ = 50 %	1/3 = 33.33 %	1/4 = 25 %	1/5 = 20 %
1/6 = 16.66 %	1/7 = 14.28 %	1/8 = 12.5 %	1/9 = 11.11 %

Simple & Compound Interest

Formula:

1) Simple Interest (SI) = P x R x T / 100

Where,

P = Principle
R = Rate of Interest
T = Time

2) Amount(A) = P + prt/100 = P (1+ rt/100)

3) Compound Interest (CI) = A – P

Where,

A = Amount
P = Principle Amount

4) A = P(1 + r/100) t

5) When we calculate ½ rate of interest of CI

$A = P (1+ r/2/100)^{2t}$

> **Note:** *We can calculate value as same on changing formula number 5 for quarter 4th T and yearly as per year specified.*

6) The Difference between Compound Interest & Simple Interest between 2-year formula.

$= P (r/100)^2$

7) The Difference between Compound Interest & Simple Interest between 3-year formula.

$= P(r/100)^2 (r/100 + 3)$

Formula Collection # 2

Sequence Series Formula

1) Arithmetic Progression (AP) = The sequence $a, a+d, a+2d, a+3d \ldots$ is known as an AP.

2) Arithmetic Mean (m) = $a+b/2$

3) Geometric Progression (GP) = $a_1, a_2, a_3, \ldots a_n$

Where, Sum of n terms (Sn) when $r < 1$

   a) $Sn = a(1-r)n/(1-r)$
   b) $Sn = a(r^n - 1)/r-1$, where $r > 1$
   c) If sum of an infinite GP $a, ar, ar2\ldots$ is $a/(1-r)$

4) Harmonic Progression (HP) = $a1, a2, a3\ldots$ is called an HP.

Where, $1/a1, 1/a2, 1/a3 \ldots 1/an$

Combination & Permutation

1) $^nP_r = n! / (n-r)!$ where r is taken time in Permutation formula.

2) If difference time taken = $^nP_n = n!$ Where $0! = 1 = 1!$

3) Combination formula

$^nC_r = n! /(n-r)! r!$, where C is used for combination sign.

4) $^nC_r = {}^nP_r / r!$

5) $^nC_0 = {}^nC_n = 1$

Probability

1) P(E) = Number of favorable outcome(Element) / number of possible outcome (Element)

Geometry

1) Area of Rectangle = l×b

Where, l = Length; b = Breadth

2) Area of Triangle = b x h2 or 1/2 x length x height

Where, b = base of the triangle; h = height

3) Area of a Circle = A = π×r2 , [Where π = 22/7]

4) Circumference of a Circle=A=2πr

Where R = radius

5) Surface Area of a Cube = 6a2

6) Volume of a Cylinder=V=πr2h   [where ,h = height]

or

π (d/2)² x h

7) Volume of a Cone = V = $\pi r^2 h$ or $1/3 \times \pi(d/2)^2 \times h$

8) Area of a Square = a^2

END

Author Bio

SUJIT KUMAR MISHRA

Sujit Kumar Mishra is the Author of the first Mysterious Island Zoya Novels. He is also Software Engineer, Actor & Author of non-fiction books *Dot net & MVC interview questions: Interview preparation*, Android Books for Practical, Android Phones & Tablets Development, Mathematics One Day and How to be rich Tricks (Hindi / English) which is published on 2019.

Instagram: @mishrasujitkr

Twitter: @sujitmi42019201

www.ingramcontent.com/pod-product-compliance
Lightning Source LLC
Chambersburg PA
CBHW050006230526
45465CB00003BB/1277